It would be rough to be a bug.

There is danger every place a bug goes!

A busy bug could walk into a web. ...

3

A frog could grab it as it flits. …

4

It could become a bat's snack. ...

A cat could swipe at it. ...

Bugs face so many dangers!

7

Bug life must be tough.